Bandboxes and Shopping Bags

in the collection of the
Cooper-Hewitt Museum

The Smithsonian Institution's
National Museum of Design

Foreword

The Smithsonian Institution has frequently and lovingly been referred to as "the nation's attic." The attic has expanded considerably in size since the Cooper-Hewitt Museum became a part of the Smithsonian family. The range and variety of the Museum's holdings never cease to amaze.

Why does the Cooper-Hewitt collect shopping bags? Early in the century the founders, Sarah and Eleanor Hewitt and their sister Mrs. James O. Green began to amass a collection of bandboxes for the Museum, and apparently encouraged their friends to do likewise. These picturesque oval paperboard boxes to carry lightweight and easily crushed accessories and linens, which have been called the shopping bags of the 19th century, were carryalls for both sexes and all levels of society, making a nice parallel with today's bags.

We are deeply indebted to the Interstate Bag Division of Champion International Corporation for providing the funds for the publication of this catalogue, which is one of the series documenting the Museum collections, as well as for a new addition to the collection, a personal Cooper-Hewitt shopping bag.

Lisa Taylor
Director

Bandbox and cover.
United States, about 1840.
Given by Mrs. Frederick F. Thompson.

The shopping bag—the child of commercial interaction in an age of consumerism—has assumed a role in contemporary life that exceeds its straightforward functional purpose. Created for the transport of relatively lightweight objects, these simple portable containers have been designed as a status symbol for the carrier, and as a graphic message about the establishment which provided it. The pedestrian carrying a bag freely provided by a chic department store, boutique, or even museum, signals to the world his or her informed and selective consumerism. At the same time, the bag is a portable extension of the commercial establishment, travelling about the city as a form of advertising. In so doing, it also has become a powerful visual presence on city streets. Usually made of paper, but sometimes of plastic or other materials, contemporary bags are emblazoned with an establishment's name, its logo, any of a number of vivid colors, striking graphic patterns, slogans, and/or messages of popular concern, to create a great variety of visual images. The shopping bag is closely associated with the pedestrian rather than the motorist. Thus, the most densely populated pedestrian areas of our cities, such as "downtown" and shopping centers, often seem to be the site of a kinetic sidewalk exhibition of graphic design, one that embodies the drama and hustle-bustle of commerce.

One antecedent of the shopping bag is the uniquely American band box, a rounded box with cover carried by both men and women. During the first half of the 19th century, the band box saw widespread use throughout the culture—it was carried by everyone, its simple function common to all levels of society. Its purpose was to contain a traveller's lightweight linen accessories, such as starched, pleated, and goffered collars, cuffs, neckbands, and house bonnets. Indeed, the name "band box" alludes especially to its role as a container for men's neckbands.

The boxes were manufactured out of cardboard, and their creation formed a lively cottage industry. They were covered, not only with available swatches of wallpaper, but more often with papers which were created specially for that purpose. However, many of the people who produced band boxes later went into the production of wallpapers. Band boxes were covered with

Bandbox.
Paper on cardboard. U.S.A., 1830. Given by the Misses Hewitt.

Bandbox, "The Old Capitol in Washington."
Printed in reds, browns and white on yellow ground. United States, about 1830. Given by Mrs. J.O. Green.

Solid colors

Halston, Ltd., New York, *paper, 12¼" x 16"*

Kron Chocolatier, New York, *paper, 9¾" x 7¾"*

Ambienti, New York, *paper, 15⅜" x 16"*

vivid and memorable images, and occasionally these were clearly advertisements. One such box, from 1823, is emblazoned with a proud eagle bearing a banderole in its beak which announces that the maker of the box is "Putnam and Roff," of Hartford, Connecticut. The images of this box are executed in a range of muted colors—light blue background, with block-printing in orange, white, gray, and light brown. The rim of its cover has the same blue background and is block-printed in white, pink, brown, and green. Other boxes depict hunters with dogs and landscapes dotted with houses and other buildings, and frequently commemorated current or political events, such as the building of the Erie Canal, Zachary Taylor, a hero of the 1846 Mexican war, or John Tyler's campaign.

Graphic design, technology, and social change have coalesced to produce the contemporary shopping bag. In the late 19th and early 20th centuries, affluent people—the carriage trade—didn't actually carry home their own packages. Servants were sent to collect purchases from a store, or the items were sent to the home by messenger. In the early decades of this century, the delivery of packages to the home became an artfully composed ceremony. The delivery van, the liveried staff and the packaging all were designed to reflect the quality-consciousness of the store. Obviously, some stores, and by extension, their delivery systems, were more prestigious then others. A new element of status revolved around which delivery truck drove up one's driveway. The status-laden image of a "carriage trade," or a select clientele, has persisted, even though the army of service people required to actually fulfill such an image has by and large disappeared. The shopping bag thus appeared in our own more egalitarian contemporary culture as a genuine need. The old-fashioned delivery boy has been replaced by a bag that embodies many of the same intimations of status. Carrying the smart bag has become the contemporary equivalent, although considerably diminished, of the smart delivery van conspicuously parked in front of the house.

The technology to produce an inexpensive shopping bag is relatively new. The bag with a handle attached cheaply and easily by machine has existed only since 1933. Before then, bagmaking involved a number of separate hand operations. For example, the handle of a bag was stapled to a piece of cardboard which was then glued to the insides of a bag.

The second technological development which allowed the bag of today to become a reality was the widespread use of the flexographic printing technique. This technique involves the use of rubber plates which are cheaper to produce than gravure plates, and which allow large areas of solid color to be printed on the uneven surface of heavy-duty paper. As recently as the mid-1930's, the designs and typography were hand-carved into rubber plates. Now, designs of any sort can be reproduced by the process of photo-etching. By 1936, one on-line machine could fabricate a complete bag and print on it any design. By the late 1930's the paper bag with a handle was sufficiently inexpensive to produce so that a store could view such an item as a "giveaway." Even with these pivotal inventions, the manufacture of paper bags remains a process that requires both the involvement of machines and of people, where the machine handle the fragile paper so gently as to seem human, and where the humans are required to be so efficient as to appear mechanized.

The memorable bags of today are also a by-product of what many people regard as a revolution in graphic design which occurred in the 1960's. All at once, or so it seemed, many ordinary products of everyday life—record jackets, book covers, menus, shopping bags, and packaging of all kinds—sprouted new and vivacious graphic qualities. The surfaces and shapes of a bag came to be regarded by the designer as an opportunity to create an exciting visual composition. No longer was it enough to print merely the name of a store on the bag. The surfaces of a bag could portray the image, tone and quality of a store, institution, or product through such various devices as the use of discreet, elegant typography, or lush, seductive, pictorial illustrations. In fact, this supposed graphic revolution of the 1960's was part of a larger phenomenon. The 20th century is really the first time that a great deal of art has been directed towards public display and for public attention and approval. This is exemplified by posters, advertising, packaging, and shopping bags.

The phenomenon occurs in two ways. One, recognized art and artists are used to attract people and as a means of advertising. Two, art is made more accessible to the public. In particular, the work of new artists is brought to public attention by having it appear in the public domain in these new forms.

The collection of the Cooper-Hewitt Museum is formed primarily with regard to this last aspect of the contemporary shopping bag, as a form of portable graphic art. Eleven categories of bag design have been noted. Of these 10 portray different approaches to graphic design, and one explores the use of unusual materials and shapes.

Many bags use a simple overall color. The color itself is the most memorable aspect of the bag—black, dove gray, vivid red, burgundy, forest green, navy blue, colors which connote a rich luxury and carefully studied and understated elegance. Often the name of the store will be displayed simply in white or a soft color. But a few bags are like Halston's hallmark red bag, and reveal no other identifying mark. These bags are the design products of an especially confident commercial establishment.

Beyond the use of a simple color, bags employ a composition of words and/or abstract elements such as geometric shapes, or lines; such composition implicitly connotes the design awareness of the store itself.

Many organizations decorate their shopping bags with only a logo, or with a combination of logo and names. A logo is a design based on typography or abstract forms which is intended to act as a succinct and familiar substitute for the more conventional proper name of an organization. The design may vary in size from quite small to one that encompasses the entire bag surface. Bags of this sort are usually designed to be part of a more larger program aimed at the coordination of all aspects of the visual image of the organization.

A logo, initial, or name can form the basis for the creation of a pattern to cover the entire surface of the bag. This ornamental use of an identifying mark is not unrelated to the use of designer initials in the creation of a patterned fabrics for bed linens, clothing and personal accessories.

Other shopping bags employ pure pattern created in a more traditional manner from repetitive abstract motifs rather than from any specific reference to the name or initials of an establishment. However, such a pattern can connote aspects of the store. The bag for Julio, for instance, makes reference to the interior setting of the shop itself, repeating on the bag the pattern of small white tiles used throughout the interior.

Some stores and museums have used their bag as a medium of pictorial storytelling about their products and exhibitions. A very famous bag, now virtually a collector's item, is the one embellished with an Andy Warhol image of a soup can. The bag, used to announce an exhibition of pop art paintings, was like one of the actual paintings.

Perhaps the most ebullient bags are those that are purely pictorial, and which emphasize a striking and fetching visual image, occasionally at the expense of a conventional verbal identification.

A related category is the use of the traditional or nostalgically old-fashioned pictorial image, which can range from a fine engraving to an example of the illustrator's art.

Some bags are amusing or tongue-in-cheek conceits. A bag with a drawing of a bag on it, or a bag pretending to be a purse is making an ironic comment on bag imagery.

The weeks leading up to Christmas are the most intense period of the year for commercial sales, and thus it is natural that stors have developed a whole range of special holiday and seasonal shopping bags as an extension of the idea of gift wrapping.

The last category includes those bags which are made of an unusual material, and/or in an unusual size. Included is a bag which is very, very small in which to carry home tiny treasures and one which is uniquely shaped to carry rolls of wallpaper. Also included are bags made of burlap, vinyl, and netting.

Richard B. Oliver
Curator of Contemporary Architecture and Design

iane Love, New York,
per, 15¾" x 15⅞"

ntitled black w. white
panese characters, Tokyo,
per, 12½" x 12½"

Gump's San Francisco, *paper, 15¾" x 15⅝"*

Andre Oliver, New York, *paper, 15½" x 16", designed by Steve Rogin*

Steuben Glass, New York, *paper, 9¾" x 7¾"*

Crouch and Fitzgerald, New York, *paper, 17¼" x 15⅞"*

Logo

Vuokko, Helsinki, Finland, *paper, 13" x 9¾"*

Royal Copenhagen Porcelain, New York, *paper, 15¼" x 13"*

Roots Shoes, New York, *paper, 13¾" x 8¾"*

Valentino, New York, *paper, 9¾" x 7¾", 10⅝" x 16", 1970*

Godiva Chocolatier, New York, *paper, 9¾" x 7¾", 1970*

Charrette, New York, *plastic, 18" x 13"*

Design Research, New York, *paper, 15½" x 13"*

Juschi, New York, *paper, 11¼" x 14¼"*, designed by Steve Lawrence with Ilene Rich

Logo as Pattern

Strawbridge & Clothier, Philadelphia, *paper, 15" x 19"*

Eastern Airlines, *paper, 16⅝" x 13"*

Marimekko, Helsinki, *plastic, 19¼" x 14⅞"*

Mark Cross, New York, *paper, 15¼" x 12¾", 1970*

Cartier, New York, *paper, 16⅝" x 7¾"*

Gimbels East, New York, *paper, 11½" x 15⅞"*

Pattern

Julio, New York, designed by Modern Art Graphics, *paper, 10¾" x 15¾"*

Estee Lauder Cosmetics, *paper, 9⅝" x 7¾"*

Lederer de Paris, New York, *paper, 19" x 13", 1970*

Delman Shoes, New York, *paper, 13¾" x 8¾"*

Azuma, New York, *paper, 13¾" x 8¾", 1970*

Haibara, Tokyo, *paper, 12½" x 12¼"*

Sample House, Dallas, Texas, *paper, 9⅝" x 7¾"*

"Campbell's Tomato Soup," Andy Warhol, Institute of Contemporary Art, Boston, *19¼" x 17"*

Conran's, New York, printed in England, *paper*, *15¼" x 12¼"*

Nikon, New York, printed in Japan, *paper, 15½" x 11¾"*

Michael C. Fina, New York, *paper, 9¾" x 7¾"*

Susan Bennis, Warren Edwards Shoes, New York, printed in Italy, *paper, 14½" x 5¾", designed by Miguel Sanchez photo by Richard L. Shaefer*

"Sundown, a joyous sunset behind two lonely weeds," The Emporium, Cheyenne, Wyoming, *plastic, 17¼" x 15¾", designed by Peter Max*

Aviance Cosmetics, *paper, 9⅝" x 7¾"*

Once Upon a Quilt, San Francisco, *paper, 9⅝" x 7¾", 15½" x 13"*

Boudin Sourdough French Bread, San Francisco, *paper, 16¾" x 13⅞"*

"Winter Festival," Aramis Cosmetics, *paper, 10½" x 16"*

Graphic

"Love is Sweet" Japan, *plastic and paper, 11¼" x 12¾". This bag was used to hold the gifts given to the guest at a wedding in 1972. Traditionally such gifts were housed in "Furoshki" (cloth) but it is not uncommon at present to see the Furoshki package housed in a paper shopping bag*

Untitled, white w. black in middle w. multi-colored Japanese symbol, Tokyo, *paper, 12" x 12¾"*

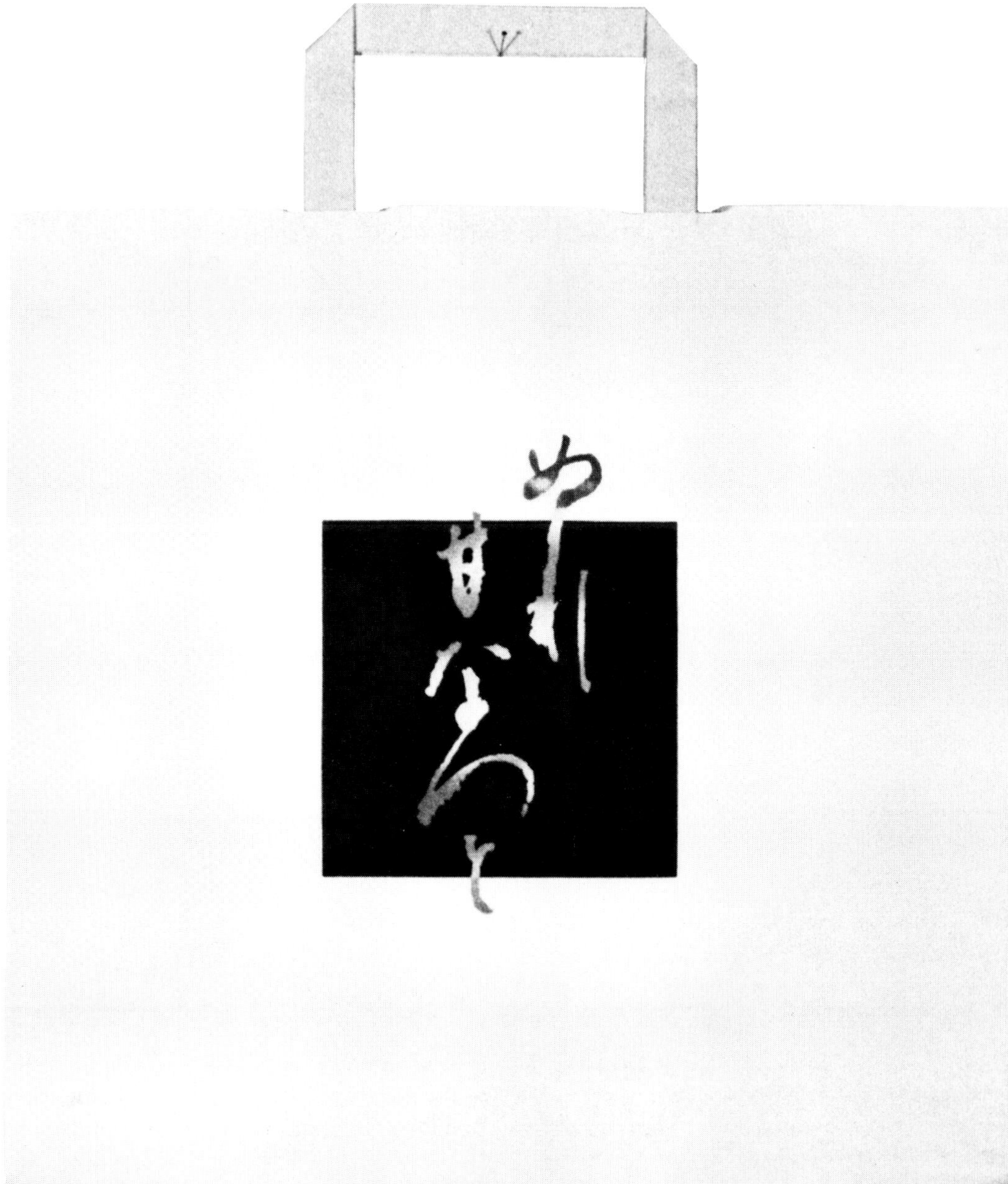

Ginza, Matsuya, Tokyo,
paper, 15½" x 12¼"

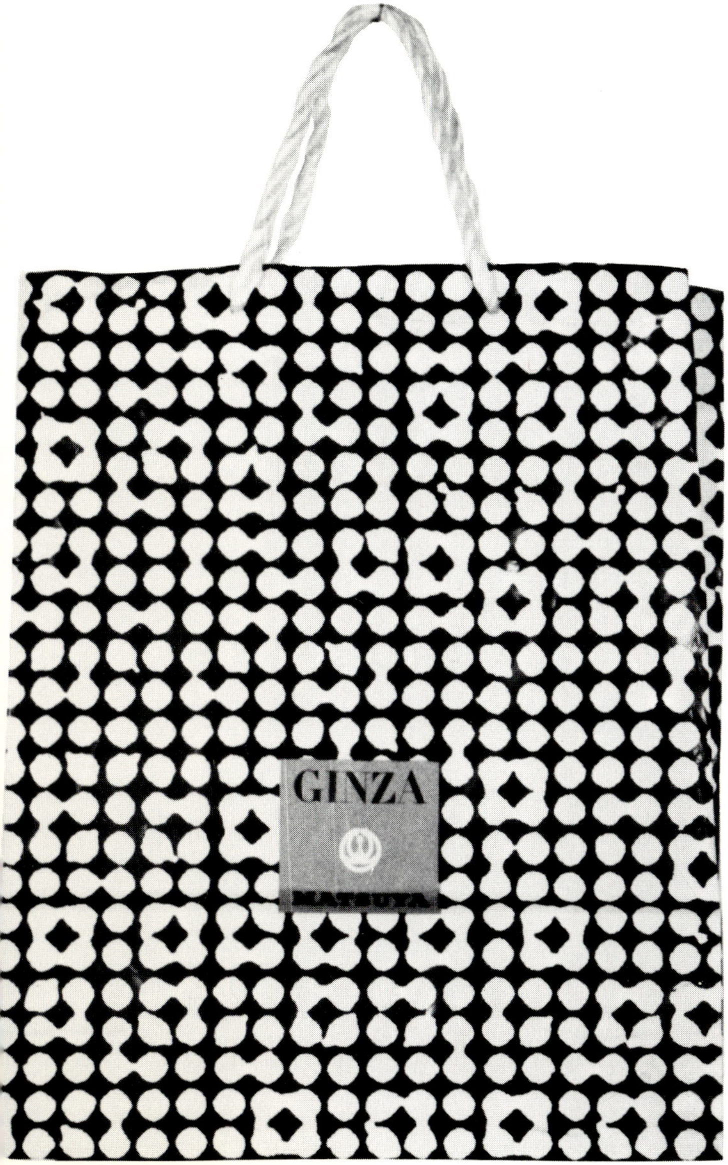

Ginza Hakubotan, Tokyo,
paper, 12¼" x 10⅞"

Graphic

"Flammarion," Librairie du Centre National d'Art et de Culture Georges Pompidou, Paris, *plastic, 16½" x 14½"*

Mario Valentino, New York, printed in italy, *paper, 17¼" x 9⅞*

London Underground Map, *plastic, 17⅞" x 14⅞", designed by Paul E. Garbutt*

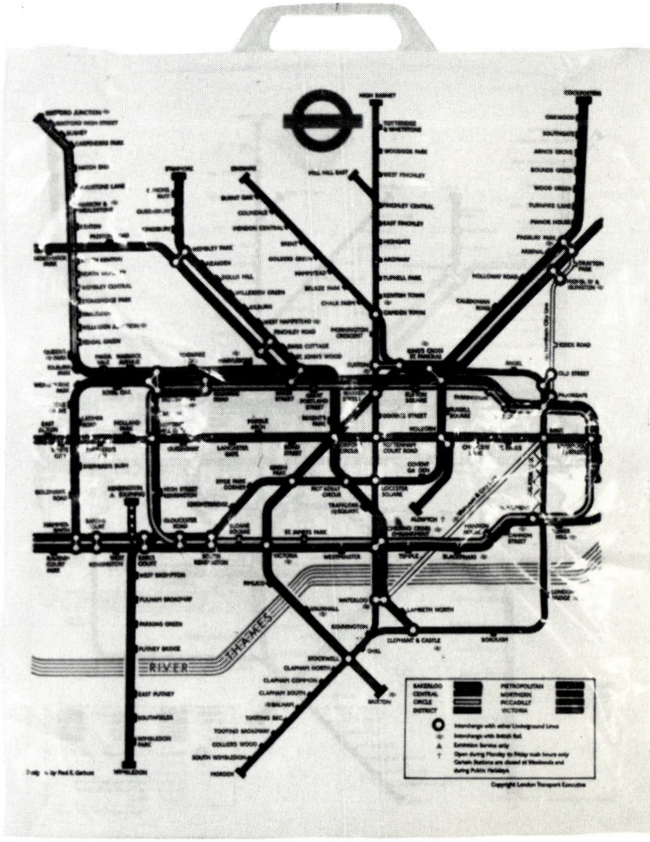

Levy's, Phoenix, Arizona, *paper, 18¾" x 17⅝"*

"Brazil Fortnight," Neiman-Marcus, Texas, *paper, 10½" x 16", 1978. This graphic reproduces Rio di Janeiro's sidewalk pattern.*

Emanuel Ungaro Parallele, New York, *paper, 10¾" x 15⅞"*

Pictorial

Bloomingdale's, New York, *paper, 16½" x 12", photograph by Marco Glaviano*

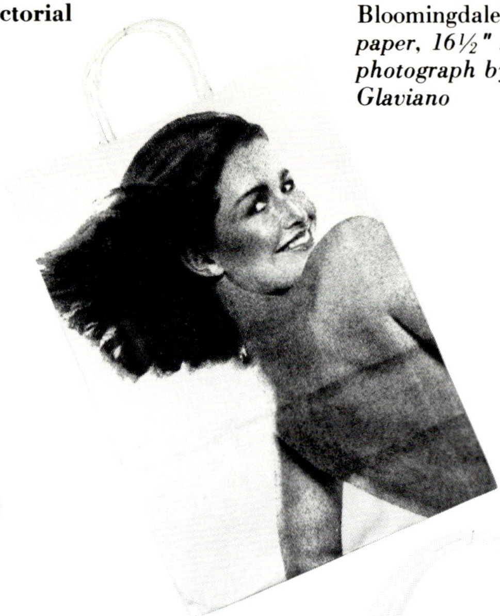

Seibu, Tokyo, *paper, 15¼" x 12½"*

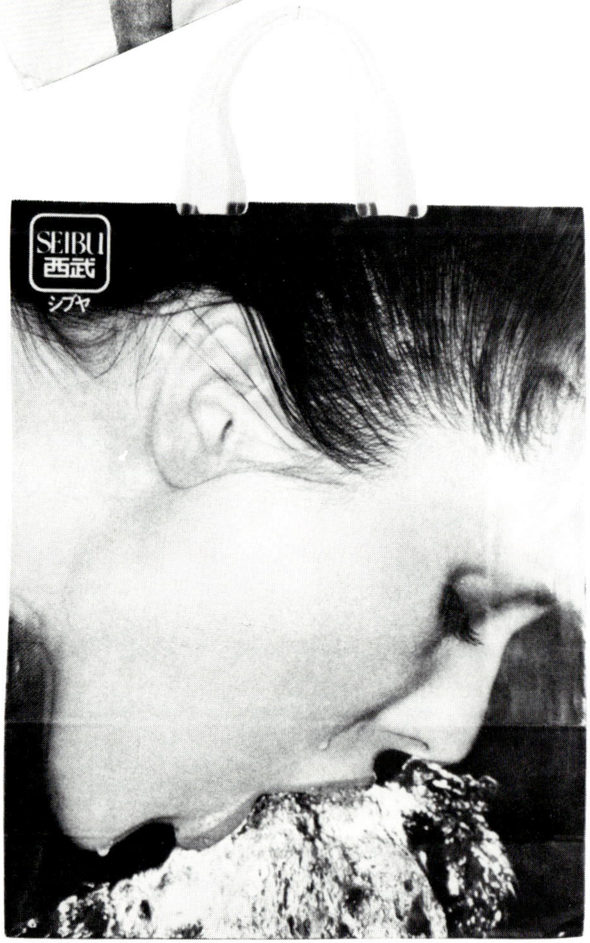

Bloomingdale's, New York, *paper, 16½" x 12", illustrated by Michaele Vollbracht, 1970*

Camouflage, New York, *paper, 14" x 11", illustrated by Mel Odom*

Bloomingdale's, New York, *paper, 16½" x 11¾", 1978, illustrated by Antonio Lopez*

Azuma, N.Y., *paper, 13½" x 8¾"*

Bernies, Calif., *paper, 15¼" x 12¾", designed by Katha Dalton*

24

Fiorucci, New York, printed in Italy, *paper, 21½" x 19½"*

Bernies, Calif., *paper,* *15¼" x 12¾"*

Fiorucci, New York, printed in Italy, *paper, 18¼" x 19⅝"*

Niagara Germany, *plastic* *16⅜" x 13⅞"*

Historic/Traditional

"United States Patent Office," *paper, 19½" x 15¾"*

Brentano's Bookshops, New York, *paper, 17¼" x 13", 1970*

McCreedy & Schreiber, New York, *paper, 17¾" x 16¼"*

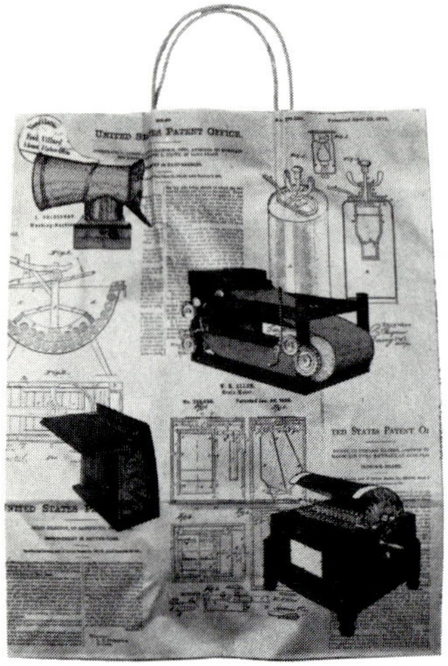

Sotheby Parke Bernet, New York, *paper, 18" x 16¼", 1970*

Taken from a poster, *Troupe de Mlle, Eglantine,* by Henri de Toulouse-Lautrec, Sherry-Lehmann, Inc., New York, *paper, 12½" x 10", 1970*

Taken from an Eighteenth century Design, *paper, 17" x 12"*

Conceit

"Tweed Jacket,"
Bloomingdale's New York,
paper, 16½" x 12", illustrated by Michaele Vollbracht, 1970

"Bag on Bag,"
Bloomingdale's, New York
paper, 16½" x 12"

"Mock Bronc Busters" Saddlebag", *paper, 8" x 7"*

Gucci, New York, *paper, 9⅞" x 8¾", 15½" x 16", 1970*

Roberta di Camerino, New York, *paper, 10⅝" x 16", 1970*

Christmas

"Christmas 1977," Neiman-Marcus, Texas, *paper, 15½" x 16"*

"Especially at Christmas," Altman's, New York, *paper, 15½" x 13", 1970*

"Christmas at Sakowitz," Sakowitz, Texas, *paper, 15½" x 16"*

"The Old Woman in the Shoe, Altman's is the Christmas Store," Altman's, New York, *paper, 16" x 13", 1970*

"Come, let us believe in magic again," Bloomingdale's, New York, *paper, 16¾" x 12", 1970*

"Illustration of Santa Claus" by Thomas Nast, from Harper's Weekly, January 1, 1881," *paper, 19⅛" x 15¾"*

"Merry Christmas," Bailey, Banks & Biddle, Philadelphia, *paper, 15¾" x 12¾"*

"Wrapped Package," Bonwit Teller, New York, *paper, 15½" x 13"*

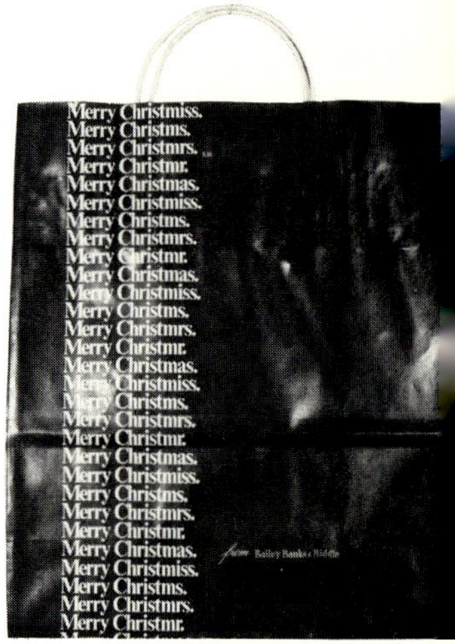

Unusual Shapes

Sonia Rykiel, Paris, *vinyl*, *13¾" x 17¼"*

Laura Ashley, New York, *plastic*, wallpaper roll bag, *12¾" x 28½"*

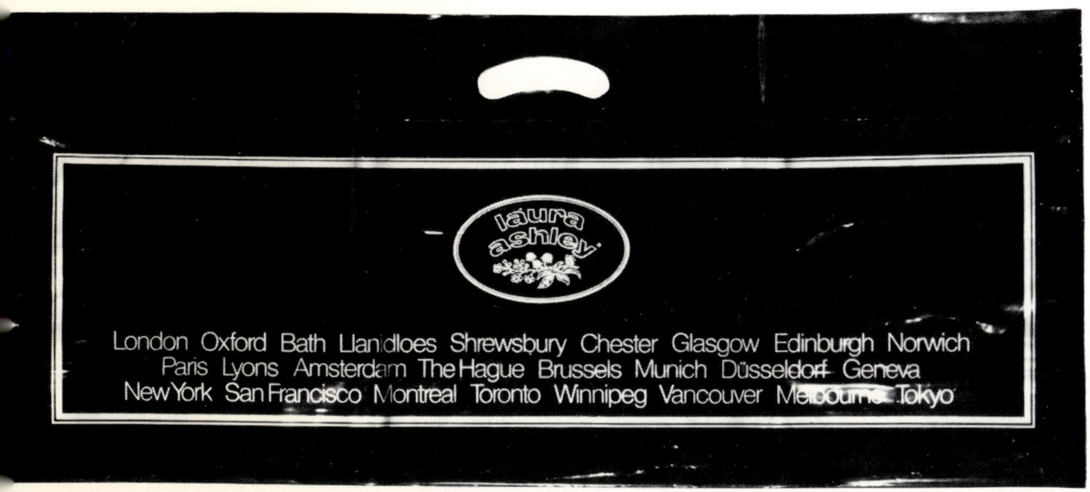

Kalso Earth Shoe, New York, *burlap*, *17¾" x 13"*

Prohibition, Paris, *paper*, *11⅛" x 5⅝"*

Hand-painted bag, India, *paper*, 4¾" x 3⅞"

Jungle Jap, Paris, *paper*, 18½" x 21"

Mesh Bag, France, *string*, Sacha, Paris, *canvas*, 15¼" x 17½"

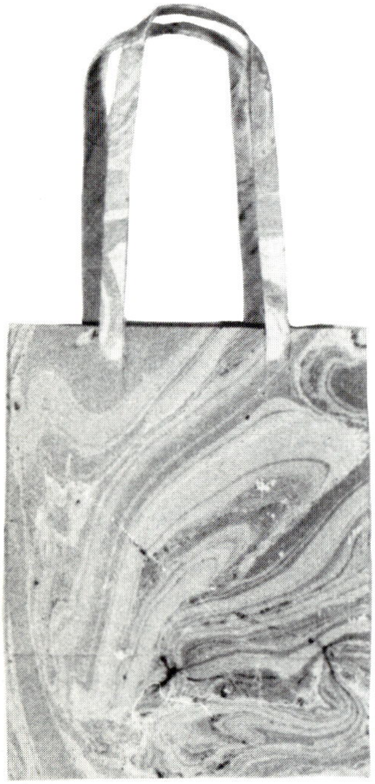